SEEKING

Angela Habbie

FIRST EDITION 2020

CONTENTS

I. Earth's Early Atmosphere
II. Oxygen
III. Requirements For LifE
IV. Looking For Habitable Planets
V. Other Worlds
VI. The Drake Equation And The Fermi Paradox

Artist impression of the Early Earth

I. Earth's Early Atmosphere

Gas bubbles trapped in ancient rocks and atmospheres of planets and moons of the solar system, have helped scientists gather data to construct theories surrounding the earth's early atmosphere.

One theory explores the idea that the early earth was carpeted with volcanoes that released carbon dioxide, water vapour and nitrogen, and that these gases formed much of the early atmosphere. Eventually, as the Earth cooled down water vapour condensed and fell as rain. Water molecules were then collected in hollows in the crust as the rocks solidified and the oceans were formed.

Meanwhile, another theory speculates that water on the earth was also brought about by icy comets as they fell onto the earth's surface and melted.

Some suggest that the mixture of gases were formed from solar debris, similar to comets, crashing into the earth and vaporising around 500 million years after its formation.

Slowly, the Earth stabilised and it is believed that the atmosphere then, mainly consisted of carbon dioxide. Water vapour, nitrogen gas and traces of ammonia and methane are also believed to have been present then.Oxygen was absent or present in traces. Present day Mars and Venus have atmospheres very similar to this.

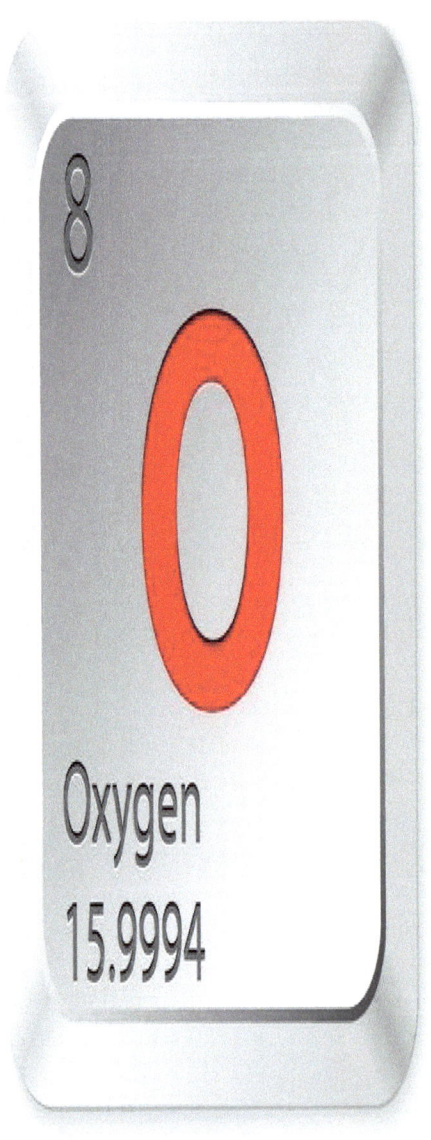

Chemical element: Oxygen

II. Oxygen

Life as we know it, would not have come about if it weren't for oxygen. Scientists believe that life began about 3.5 billion years ago, when simple organisms similar to bacteria (stromatolites) appeared. These could produce energy by breaking down chemicals.

Later, about 2.7 billion years ago, bacteria and other simple organisms such as algae evolved. They used energy from the sun to photosynthesis and produce their own food.

Oxygen is a waste product of photosynthesis and thus oxygen came about on earth. Algae and bacteria thrived in seas and eventually, the levels of oxygen increased, such that, plants evolved, who carried out the process of synthesis too- using up carbon dioxide and releasing oxygen.

(energy from sunlight)
Carbon dioxide + water → → → → → → →
→ glucose + oxygen

Plants evolved to colonise most of the surface of the Earth. Hence, the atmosphere became rich in oxygen, making it possible for the first animal to evolve. They relied on plants and algae for their food and used oxygen to carry out the process of respiration.

Several early microorganisms that had evolved in the absence of oxygen could not tolerate these high concentrations of oxygen and died as there were fewer favourable habitats that supported their survival.

In the 1950s, Harold Urey and Stanley Miller conducted an experiment based on the hypothesis that this primitive atmosphere, combined with the liquid water environment, is how life started. In a simple laboratory experiment, they exposed ammonia, methane, hydrogen, and water to an electrical discharge to kick-start any possible reaction.

It is believed that realy, ultraviolet light from the sun was the catalyst for the reaction. But at the time UV radiation was very hard to produce in a lab. After a week, the brown substance in their beaker was analysed, and was found to contain amino acids. This experiment demonstrated the possibility for life to have been present once the planet cooled enough for Earth to support liquid water as a reaction medium.

Algal bloom

III. Requirements For Life

Since we have no other examples of life, we have to model our assumptions on our own existence. These requirements are one, the presence of carbon, being able to form complex chemistry. Two, the presence of liquid water as a necessary medium for chemical reactions creating life. And three, the correct environment. For the correct environment scientists often use the term 'habitable zone', in which the proposed planet must lie within. Much of the assumptions going into the habitable zone are based on the star the planet orbits, the type of star-- longevity, type of radiation given off-- and the atmosphere of the planet-- composition, greenhouse effect.

These assumptions are well founded on logic and reasoning based on the evolution of life as we know it here on Earth. Carbon, hydrogen, nitrogen and oxygen have played a crucial part in the creation of life on earth and seem to be a necessity for life. 'Extremophiles' however are known to exist in extreme conditions- conditions that one would normally expect to not support life. If extremophiles can survive here on earth in an environment lacking these essential ingredients for life, surely planets that have different environments to earth can support a different form of life that has adapted to thrive in that planet's environment.

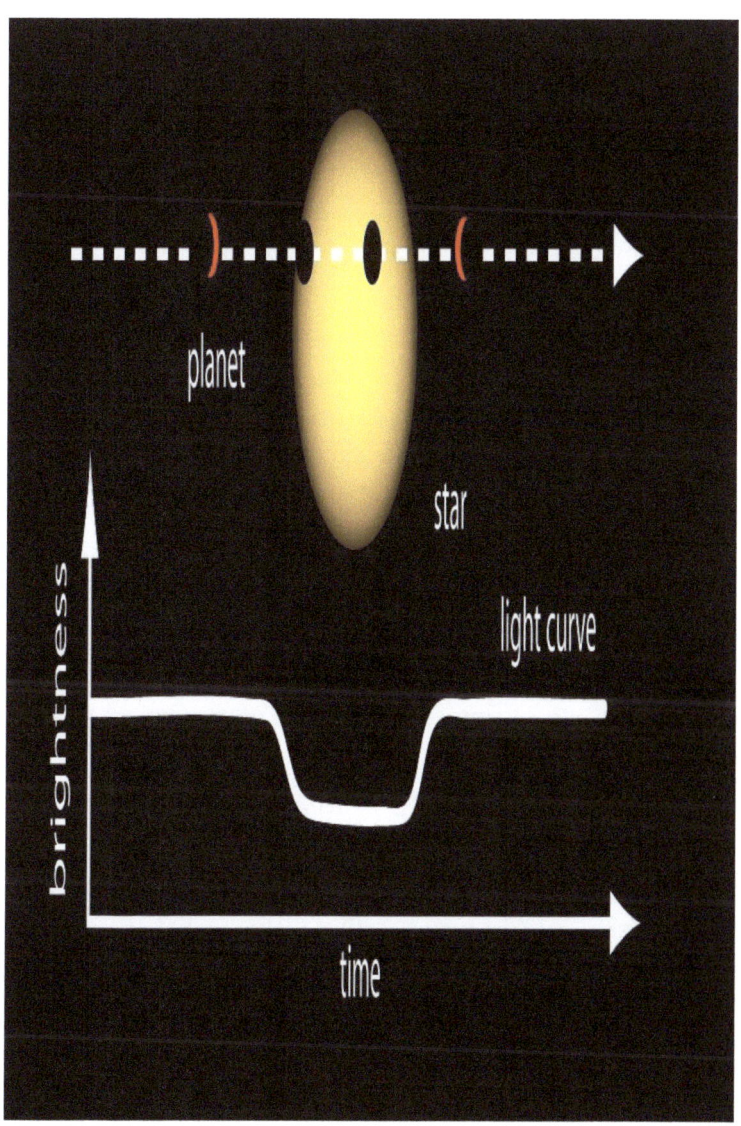

Light curve of a planet transiting its star

IV. Looking For Habitable Planets

When searching for the existence of a habitable planet beyond our solar system, it is important to know where to look. A planet that orbits a star outside of our own solar system is referred to as an 'exoplanet'. They are detected using transit spectroscopy. For this method, astronomers monitor stars for dips in their luminosity, which are the result of planets passing in front of the star relative to the observer.

A recent estimate is that, on average, there is at least one planet around every star in the galaxy. Rather than pointing a telescope at every star in the sky and getting potentially very few results, scientists have developed models to understand where the greatest probability of finding a habitable planet would be.

So, what do you think you would investigate first, when looking for a new star-planet system to study? What other conditions might improve the potential habitability of a planet within a star-planet system? Let's break it down. There are a few major factors to consider when assessing the probability of finding a habitable planet:

1. Type of star that it's orbiting (its temperature, single or double star system, preferably with no extreme flaring, a O, B, A, F, G, K or M type star).

2. Distance between the star and planet.
3. Size and other characteristics of the planet (eg. its density, strong magnetic field to shield it from solar radiation).
4. Age of the system.
5. Signs of biological activity(eg. The presence of elements like oxygen).

Stars similar to the Sun and long-lived dwarf stars with planets of comparable size and orbit to Earth's would be most likely to sustain life as we know it. The presence of elements which compose life we are familiar with, including carbon, hydrogen and nitrogen, would also increase the likelihood of habitability on a planet. Furthermore, according to NASA, planets on which the chance of life existing is greatest are small and largely rocky, based upon current knowledge of Earth's ability to sustain life due to its characteristics. So, for a while now, scientists have been searching for exoplanets located in their star's habitable zone (the distance from a star at which a planet is warm enough to have liquid water on its surface and thus potentially support life as we know it on Earth) in order to understand the Earth's evolution.

Trappist-1 System

V. Other Worlds

For many years, humans have questioned the possibility of the existence of life on other celestial bodies and finding systems with features similar to that of the Earth. A recent discovery in this field has the entire scientific community electrified.

The TRAPPIST-1 System

In 2016, astronomers at the Transiting Planets and Planetesimals Small Telescope (TRAPPIST) in Chile identified three planets around the dim star TRAPPIST-1. Less than a year later, NASA announced the discovery of even more worlds, for a total of seven. All of the exoplanets orbit in the habitable zone of their star, the region where water can remain liquid at the surface. The TRAPPIST-1 system boasts the largest number of rocky worlds ever found in a habitable zone of a single star and lies only 40 light-years from Earth, as part of the Aquarius constellations

The TRAPPIST-1 planets are easy to observe and thus characterize with telescopes because they orbit in sync; together, the seven exoplanets form a resonance chain (they are tidally locked) connecting them all together and suggesting a slow, peaceful evolution. The planets are transiting (frequently) in front of their star, The planets are also a lot closer to their star, than Earth is to the Sun.

So, why are the planets thought to be capable of supporting life in the Trappist-1 system, if the planets are a lot closer to their parent star than Earth to ours?

In the TRAPPIST-1 system, the star is classified as an M-type, ultra-cool red dwarf with an extremely small radius, which would indicate that the temperatures on planets in close orbit to TRAPPIST-1 are cooler than those expected at planets with the same orbital radius in the solar system. Therefore, according to some studies, liquid water would be able to survive on planets relatively close to the star, more so than in the solar system. Researchers have found densities of the worlds ranging from 0.6 to 1.0 times Earth's density. The seven worlds are rich in water, with water levels on some reaching as high as 5 percent of the total mass. In comparison, only about 0.02 percent of Earth's mass is contained in water. Trappist-1 may be too wet for life, however some scientists suggest it is life friendly. On top of that, there's the nature of red dwarf stars, which are prone to flare-ups that could wreak havoc on their planets' atmospheres.
However, studies have found that exoplanets orbiting red dwarfs could still be habitable as long as they had sufficient atmospheres and cloud cover to deal with the radiation.

The Seven Exoplanets in the TRAPPIST-1 System

TRAPPIST-1b and c, the innermost worlds, are likely to have rocky cores and be surrounded by dense atmospheres thicker than Earth's. Lying close to their star, the hottest worlds probably have thick, steamy atmospheres, while the most distant ones could be covered in ice. TRAPPIST-1b is the closest planet to the star, and is too hot for even sulfuric-acid clouds of Venus to form.

TRAPPIST-1d is the lightest of the seven planets, weighing about 30 percent of Earth's mass. Its low mass could be caused by a large atmosphere, an ocean or a frozen icy layer.

TRAPPIST-1f, g and **h** lie far enough from their host star that water could be frozen into ice across their surfaces. The thin atmospheres would probably lack the heavier molecules found on Earth.

Then there's **TRAPPIST-1e**, the most Earth-like of the group. As the only planet slightly denser than Earth, TRAPPIST-1e likely has a denser iron core, and may lack a thick atmosphere, ocean or ice layer. It lies in its star's habitable zone. This planet may also have a lot of oxygen, the researchers said. As water evaporates from a planet's surface and is zapped by a star's ultraviolet emissions, the hydrogen and oxygen molecules split, according to the statement. The hydrogen can be light enough to leave the atmosphere despite the planet's gravity, but oxygen will linger. Therefore, TRAPPIST-1 e could have a thick oxygen atmosphere formed through processes that are far different from anything yet observed, according to researchers.

"This may be possible if these planets had more water initially than Earth, Venus or Mars," Andre Lincowski, a doctoral student at the University of Washington and lead author of a paper published Nov. 1 2018 in Astrophysical Journal. said. "If planet TRAPPIST-1e did not lose all of its water during this phase, today it could be a water world, completely covered by a global ocean. In this case, it could have a climate similar to Earth."

Two other exoplanets of the TRAPPIST-1 system also seem to reside in the habitable zone. TRAPPIST-1d rides the inner edge of that zone, while further out, TRAPPIST-1h, orbits just past the outer edge of this "Goldilocks-like" region, according to the statement. "This is a whole sequence of planets that can give us insight into the evolution of planets, in particular around a star that's very different from ours, with different light coming off of it," Lincowski said. "M dwarf stars are very different, so you really have to think about the chemical effects on the atmosphere(s) and how that chemistry affects the climate."

Models of the star's radiation and chemistry demonstrate the possible spectral signatures of gases in its atmosphere.

It has been determined that the seven TRAPPIST-1 planets are unlikely to have hydrogen-dominated atmospheres. It has also been suggested that the atmospheres (if present) of the TRAPPIST-1 planets are most likely to be carbon dioxide-dominated, oxygen-dominated or water-dominated.

In other words, of the seven TRAPPIST-1 planets, those that have atmospheres are likely to have the kind that are favorable to life (at least as we know it). That means carbon dioxide, an essential climate stabilizer necessary for photosynthetic organisms, oxygen gas, nitrogen, and volatile elements like water. It also includes cloud cover, which is not only an indication of water, but provides protection against stellar radiation.

Researchers hope that next-generation missions—in particular the James Webb Space Telescope (which is expected to launch in 2021) and the near-infrared ground-based spectrographs used in combination with the Spitzer, Hubble and Kepler telescopes—will have the power to detect 'heavy' molecules such as **carbon dioxide**, oxygen, methane, etc. and thus they may have the potential to determine whether or not the

TRAPPIST-1 **planets** have atmospheres, and if so, what they are made of.

The Milky Way Galaxy

VI. The Drake Equation And The Fermi Paradox

In 1961, astrophysicist Frank Drake developed an equation to estimate the number of advanced civilizations likely to exist in the Milky Way galaxy. The equation has proven to be a durable framework for research, and space technology has advanced scientists' knowledge of several variables. However, the variables can only be guessed, for example, L, the probable longevity of other advanced civilizations. The equation is as follows:

$$N = R_* f_p n_e f_l f_i f_c L.$$

N = The number of technologically advanced civilizations in the Milky Way galaxy whose electromagnetic emissions are detectable.

R* = The rate of formation of stars in the galaxy.

fp = The fraction of those stars with planetary systems.

ne = The number of planets, per solar system, with an environment suitable for life.

fl = The fraction of suitable planets on which life actually appears.

fi = The fraction of life bearing planets on which intelligent life emerges.

fc = The fraction of civilizations that develop a technology that releases detectable signs of their existence into space.

L = The length of time such civilizations release detectable signals into space.

Over the years, scientists have attempted to tweak the variables of the original equation to answer more than one line of inquiry. One such example is the 'Archeological form' which eliminates L and aims to answer the question of the likelihood of there ever having been a civilization such as ours to exist rather than calculating the number of technologically advanced civilisations to exist. The equation is as follows:

$$A = N_{ast} * f_{bt}$$

N_{ast} – the number of habitable planets in a given volume of the Universe

f_{bt} – the likelihood of a technological species arising on one of these planets. The volume considered could be, for example, the entire Universe, or just our Galaxy.

The difference between the two equations is that the original one aims to find the number of technologically advanced civilizations in the Milky Way galaxy (whose electromagnetic emissions are detectable) while the other; the number of technological species that have formed over the history of the observable universe.

While scientists do not yet know the magnitude of the likelihood that an intelligent technological species will evolve on a given habitable planet, the archaeological method allows them to tell exactly how low that probability would have to be for humans to be the only civilization that the Universe has produced. It asks us to think: 'Are we the only technological species that has ever arisen? And if not, where is everyone?'- a question first put forth by Enrico Fermi, that has led to the emergence of the Fermi Paradox. The possibilities of answers to the latter are endless, and speculation will no doubt continue forever, or until we find extraterrestrials in the 'habitable zones' or elsewhere. Regardless, we can conclude that the Earth, its existence, history and processes are unique, and many, still a mystery to us. We may solve them one day or maybe never, yet we continue to search and perhaps so do 'they'. We are all seeking Earth .

THE END

References

https://www.nasa.gov/

https://phys.org/

https://www.space.com/

https://www.expertguidance.co.uk/new9-1-aqa-gcse-chemistry-c13-the-earths-atmosphere-kerboodle-answers/

https://wdet.org/posts/2019/11/07/88835-this-summers-algal-bloom-in-lake-erie-was-large-but-could-have-been-worse/

https://www.discovermagazine.com/the-sciences/how-big-is-the-milky-way

https://www.livescience.com/28738-oxygen.html

https://www.airspacemag.com/daily-planet/what-was-it-really-early-earth-180974176/

https://www.nasa.gov/mission_pages/kepler/multimedia/images/transit-light-curve.html

https://www.nasa.gov/feature/jpl/new-clues-to-trappist-1-planet-compositions-atmospheres

ABOUT THE AUTHOR

Angela Habbie has co-authored the books Fairytales 2020 & Baby Goes to Space.

www.ingramcontent.com/pod-product-compliance
Lightning Source LLC
Chambersburg PA
CBHW041944240526
45473CB00033B/504